# MATHEMATICS IS THE POETRY OF SCIENCE

*Cédric Villani*

Translated by
*Malcolm DeBevoise*

Illustrations by
*Étienne Lécroart*

**OXFORD**
UNIVERSITY PRESS

# OXFORD
UNIVERSITY PRESS

Great Clarendon Street, Oxford, OX2 6DP,
United Kingdom

Oxford University Press is a department of the University of Oxford.
It furthers the University's objective of excellence in research, scholarship,
and education by publishing worldwide. Oxford is a registered trade mark of
Oxford University Press in the UK and in certain other countries

First Edition published in 2020

Impression: 1

Published in the United States of America by Oxford University Press
198 Madison Avenue, New York, NY 10016, United States of America

British Library Cataloguing in Publication Data

Data available

Library of Congress Control Number: 2019953022

ISBN 978–0–19–884643–7

Printed and bound by
CPI Group (UK) Ltd, Croydon, CR0 4YY

# MATHEMATICS IS THE POETRY
# OF SCIENCE

# Foreword

Elisa Brune

What relationship does science bear to literature, and, more particularly, mathematics to poetry? At first sight, the answer is simple: none at all! It is often said that science disenchants the world, empties it of its poetry; that it puts turbines and machines where once there had been elves and fairies. Wasn't the moon more beautiful before human beings walked on its surface and left their landing equipment behind? Let me try to reply to the question by briefly considering the various arguments that can be made for and against the existence of such a relationship.

## The Arguments Against

Science is not poetic, for it rests on a certain conception of progress, the idea that knowledge advances by building on prior results, which are then gradually forgotten or at least deprived of their former prominence. Art, by contrast, is an unending landscape of small islands of more or less equal extent and importance. On the one hand, there is the makeshift ladder that scientists go on indefatigably constructing for themselves, each one fighting to get on top of the shoulders of the one before him, and, on the other, the timeless archipelago where artists have always lived in peace with one another and always will.

A second argument concerns refutability. One of the basic tenets of science is that all claims must be disprovable. In principle, anything that may be asserted is liable to be disconfirmed by experiment. In art, such a notion has no place. A work of art is neither true nor false; it is irrefutable.

There is also the idea that the discovery of a natural phenomenon does not really depend on the scientist who first draws attention to it. Electricity, if it had not been discovered by Gilbert, would have been discovered sooner or later by someone else; so too radioactivity, if it had not been discovered by Becquerel, and so on. It often occurs, moreover, that discoveries are made simultaneously in different places. The personal histories of scientists therefore do not occupy a central place in scientific investigation. Among artists, the opposite is true. No one other than Beethoven could have composed the *Ode to Joy;* no one other than Picasso could have painted *Guernica.* Whereas art is profoundly individual, science is a collective enterprise.

Still another argument involves the status of experiment. Art and science both conduct experiments, but in ways that differ in a crucial respect. The scientist introduces a system of measurement while standing apart from what is being measured. The artist, by contrast, is part of the experiment that he carries out; he includes himself in what he observes.

Finally, there is the dependence of science on technology, since it advances by means of increasingly precise measurements that constantly require new measuring instruments. The artist, for his part, freely chooses from among existing techniques. Great works of art can still be made using pencil and paper.

## The Arguments For

First, scientists see themselves as creators. "I have spent forty years of my life trying to be creative in physics," the French Nobel laureate Pierre-Gilles de Gennes remarked. "I am told that there is no creation in the sciences, but in that case my life has been pointless."[1] Even allowing for the doubtful chance that this is an idiosyncratic opinion, if one goes back in time one finds that the distinction between art and science grows less and less clear. It must be kept in mind that originality in art is a relatively recent invention. There was a time not so very long ago that artists did not sign their works, a time when they were interested in imitating models, not in creating something new. They sought to make identical copies, to perpetuate a tradition without adding anything to it. Little by little the artist's personality came to be given expression in a work of art.

Moreover, the arts and sciences issued from a common source in antiquity. Before they became sciences, medicine and astronomy and botany formed a whole with music and sculpture and architecture. The Greek philosophers brought order to this heterodox corpus by associating different activities with different purposes and laying down criteria of excellence, with the result that the new disciplines began to branch off from one another. Those disciplines that relied on concepts and rules of reasoning became sciences. Yet poetic intuition continued to have a place in them, for it often contains the seeds of theoretical research and scientific explanation. In many mythical and poetic accounts, one finds the outlines of an explanation that was to be more rigorously formulated later. Ovid and Darwin, for example, are talking at bottom about the same thing. Ovid describes it in a picturesque manner, Darwin in a precise fashion, but at the heart of each of these two very different perspectives one finds the idea of metamorphosis as the motive force of the world.

Another argument is that the artist and the scientist are animated by the same fundamental impulse, an intense desire to know, a *libido sciendi*, as it might well be called; and that this passion literally inhabits them, sometimes filling their lives to the point of preventing them from keeping their various interests separate from one another. Everything is an object of fascination for them, everything is grist for their mill. This is something quite different from the operation of pure reason; here reason and emotion go hand in hand. "Ardor animates reason," the great physiologist Claude Bernard used to say, "and reason guides ardor."[2]

The artist and the scientist also share a common source of inspiration: a sense of wonder, of awe before the marvels of nature, that may be said to be poetic in itself, or at least capable of provoking a poetic emotion. "The pleasure of watching water in bathtubs or mud puddles on sidewalks," as Richard Feynman once observed, "that is what makes every child a physicist."[3] A poet would have said the same thing, only with this difference—it is what makes every child a poet. Each of them, the poet and the scientist, retains the capacity for astonishment in an age when most people have learned no longer to be astonished by anything. Some would say that the curiosity of the scientist has more of the spirit of enchantment about it than that of the poet. Feynman himself was certainly of this opinion: "What men are poets," he asked, "who can speak of Jupiter as if he were a man, but if he is an immense spinning mass of methane and ammonia must remain silent?" For is not the immense spinning mass something far more extraordinary? In the same way, why should we deplore the fact that men have walked on the Moon, and regret that a part of our world has been somehow disenthralled? Traveling to the Moon made it possible to send probes to Jupiter, Neptune, and the other planets, and from there to venture beyond the Solar System. Today we are in the process of cataloguing hundreds of extrasolar planets whose existence no one had even suspected a century ago.

The ability to be amazed by nature is therefore a condition of doing important work in both art and science, though it assumes rather different forms. One may take nature either as an object or as a model, which is to say as something to be either worked on or worked with. This is what Picasso meant when he said, "One must not imitate nature, in the sense of reproducing or copying it; one must act as it does." One must, in other words, as he also put it, grow one's own branches. The scientist, for his part, takes nature as an object; he studies it from the outside, but he needs to have the temperament of an artist. On this point Baudelaire and

Einstein were of the same mind: for Baudelaire, "The imagination is the most scientific of the faculties"; for Einstein, "Imagination is more important than knowledge. For knowledge is limited, whereas imagination embraces the entire world, stimulating progress, giving birth to evolution."

Imagination and a sense of wonder are therefore inseparable companions. But there is also the role played by abstraction, that is, the selection of certain aspects of reality that one wishes to examine closely, apart from from everything else. This is the first step in undertaking either an artistic or a scientific project. In order to understand what a problem involves, in order to be able to work on it, it must be isolated and illuminated. Only then can the effort of analyzing and interpreting begin, and with it the formulation of a scientific theory or the creation of a work of art. To cite Einstein once more, science consists in extracting from a mass of sensory data a certain amount of pertinent information, and then in placing it in correspondence with concepts that, in going beyond sensations, stand alone as free creations. For the scientist, concepts are what allow him to create, the plastic matter that he shapes and sculpts.

Considering all these things, then, it will be clear that we are dealing with two activities that grow in different directions out of a common substrate. The risk of concentrating on one to the exclusion of the other will be apparent as well: excessive rationalism on the scientific side, sterile intuition on the poetic side. Henri Michaux frequently used the language of science in his writings as a way of combating the wooliness of the poets of his time, whose work, at its worst, he saw as a clutter of mere impressions from which there was nothing to be learned.[4] To his way of thinking, an artist must be able to draw upon both sensibilities in order to strike the right balance between precision and vagueness.

We need to keep in mind, too, that the arts and sciences are undergoing a radical transformation today. Since the early twentieth century, and particularly with the advent of quantum mechanics, a substantial part of the sciences' foundations has been undermined by uncertainty. When one no longer knows whether a cat is dead or alive, when one no longer knows whether a particle is here or there, or in both places at once, many previously unimagined things become possible. Not that the sciences are becoming less rigorous, far from it; but their purpose has had to be reconceived. Science, as the French philosopher Michel Bitbol has remarked, is now more concerned with investigating new directions in research than with assuring the immediate effectiveness of new discoveries.[5] In much the same spirit, André Breton prophesied more than half a century ago that "the day will come when scientists approach their work in a poetic spirit, a spirit that seems at first sight to be so foreign to them. Are we free at least to some small extent? Will we travel this road to its end?"[6]

Artists, for their part, not content to be fascinated by science, even to the point sometimes of behaving as scientists do, have increasingly adopted a theoretical point of view. This perspective is now bound up with what it means to be an artist, placing one's work in the context of a theory and commenting upon it. Scientists, in turn, have increasingly become actors in their own research, as both observers and that which is observed. Science has become a part of its own subject matter, whether in quantum physics, with its entanglements and paradoxes of measurement, or in cognitive science, which inquires into the workings of the mind. The scientific mind, in other words, now studies itself.

The scope of the physical sciences is limited by comparison with the infinite complexity of the Universe because they operate, as we have seen, on the basis of notions of refutability and of measurement. Unavoidably, because only that which is measurable can be measured, they are bound to consider only a part of reality. This may not be true of mathematics, however, for while it likewise operates on the basis of rules and reasoned argument, it is not always obliged to agree with the contingent facts of the physical world. Often it does, of course, insofar as it lends itself to the description of natural phenomena; but it may also frolic with complete abandon in spaces of thirty-six dimensions and in the vast realm of imaginary numbers, without anyone asking it to reply to a laboratory experiment. I would therefore say that mathematics is the freest of the sciences. Cédric Villani, I suspect, may not wholly agree.

Translated by Malcolm DeBevoise

## Endnotes

1. The French physicist Pierre-Gilles de Gennes (1932–2007) was awarded the Nobel Prize in 1991 for his work on liquid crystals and polymers.
2. Claude Bernard (1813–78), a French physician and physiologist, is considered one of the founders of experimental medicine.
3. Richard P. Feynman (1918–88) was an American physicist known for his work on quantum electrodynamics, quarks, and superfluid helium, as well as for his many books of popular science. Together with Sin-Itiro Tomonaga and Julian Schwinger he received the Nobel Prize in 1965 for his work on quantum electrodynamics.
4. Henri Michaux (1899–1984) was a Belgian-born French writer, poet, and painter. Apart from his purely poetical writings, he composed notebooks of real and imaginary travels and accounts of his experiences with drugs, particularly mescalin and cannabis.
5. Michel Bitbol (b. 1954) is a French philosopher of science specializing in quantum mechanics and quantum field theory.
6. André Breton (1896–1966) was a French writer and poet remembered chiefly for his role in founding and developing the theory of Surrealism.

# Contents

# 1

## Mathematics, Science, and Poetry

Mathematics, whatever else it may be, is a science.* There are those
who like to say, for a variety of reasons, that mathematics stands

* Rather than speak of *les mathématiques*, as is customarily done in French,
I decided to adopt the singular form, *la mathématique*, when I realized that I could
give no justification for the plural, unless it is an archaism recalling the Platonist
classification of arts and sciences. Moreover, there is an ancient tradition of using
the singular—carried on in our own time by Bourbaki, for example—that
insists on the unity of a discipline whose branches, though they are quite varied,
rest on common principles. I have nonetheless retained the plural noun and
verb in Léopold Sédar Senghor's fine phrase, "Les mathématiques sont la poésie

apart, but I am one of those who say that mathematics is first and foremost a science. Like all the sciences, it seeks to describe the world, to understand the world, to act on the world. Describing, understanding, acting—thus the holy trinity of the sciences.

By themselves, these three things do not suffice to uniquely characterize scientific investigation, for they are found in other activities as well. One must also take into account certain fundamental principles that are common to mathematics and the other sciences.

The first of these principles is a priori skepticism. In science, only a logical chain of reasoning can lead us to believe something to be true, not appeal to some higher authority, whether a person, or a sacred text, or something else. One must not believe anything unless one has been convinced by a rigorous and coherent argument.

The second principle is peer review, the submission of the results of one's research to the judgment of a community of experts. A result is true not because somebody asserts it to be true, but because one's colleagues have unanimously approved the reasoning adduced in support of it. This review process is seldom straight-forward. Yet in spite of disagreement, controversy, and occasional errors of judgment, scientists have always insisted upon the sharing of information and validation by qualified referees.

Third, and finally, there is the principle that no one's word counts for more than anyone else's. Only exactitude, the precision of the arguments advanced, and the conviction they inspire can lead to general agreement. In practice, of course, one more readily trusts the opinion of a respected scientist than an unknown amateur. If an article claiming to prove a difficult theorem in algebra appears under the name of Jean-Pierre Serre,[1] for example, one is naturally

des sciences," from which the present book takes its title. [In English, *mathematics* is plural in form but singular in construction, and so the problem does not arise, or at least not in so stark a form.—Trans.]

inclined to give it greater credence than if it comes from a mathematician no one has heard of. But it is important to keep in mind that here we are dealing with a human failing, something that falls short of the scientific ideal. In principle, the famous mathematician and the obscure mathematician should be considered as equals; and in fact it sometimes happens that an obscure mathematician working alone, or very nearly so, solves an important problem that has long stymied the greatest experts (this was the case only a few years ago with Yitang Zhang, and in the late 1970s with Roger Apéry; both were about sixty years old when they made discoveries that were to bring them worldwide fame).[2]

Skepticism, reasoned argument, sharing of results, peer review, collegial respect—all these things are found in mathematics, and this is why it can be said to be a science. Indeed, they are taken to extremes in mathematics. In mathematics, one is authorized to believe something only once it has been given a complete demonstration. A mathematician does not say, "Given this, you can imagine that by analogy," but rather, "I am going to prove this to you, down to the very last detail, and the logical force of my reasoning will compel your conviction." Here again, of course, we are talking about an ideal that is not humanly attainable: every proof contains small gaps, small omissions, but in principle an argument can be thoroughly verified, step by step, on the basis of the original article. Some proofs are hundreds of pages long; some take years to be verified.

Now let us consider the class of conjectures, which is to say statements that are thought to be true even though for the moment no proof can be given. Every day conjectures are finally proven (or disproven); every day new conjectures are advanced. But some are more memorable than others; Goldbach's conjecture and Collatz's conjecture (the Syracuse problem) are among the most famous of them. Mathematicians agree that Riemann's hypothesis is *the* most famous of all. It has been verified for ten thousand billion values, without a single counterexample having been encountered; but in the eyes of mathematicians, ten thousand billion corroborating instances, even a hundred thousand billion, do not constitute a proof! No other field of knowledge is so demanding.

Allow me now to qualify, for a third time, what I have just said. On the one hand, in all mathematical applications, one constantly takes the liberty of relying on statements that have not been rigorously demonstrated, but that are thought to be true because they are supported by a combination of reasoning and experiment. On the other hand, mathematicians themselves are often inclined to believe in a conjecture if it has been verified to a very large extent: even though they do not consider it to be established, they feel sure that it must be true, and this often is enough for them to feel justified in acting as though it had, in fact, been demonstrated. And while it is indeed deductive reasoning that decides whether a mathematical truth can be considered to be established, it is also true that mathematicians constantly make use of inductive reasoning and thought experiments in order to catch a first tentative glimpse of the results they seek to prove. Even when all this is taken into account, however, there is no denying that mathematics, more than any other science, enforces exceedingly stringent requirements in the way of rigorous proof.

One might go so far as to see mathematics as the quintessence of science, not in the sense that it is superior to the other sciences (even if Einstein once claimed as much),[3] but in the sense that the respect it shows for the principles of scientific reasoning is unsurpassed. In mathematical arguments, rigor enjoys a pre-eminence that would be considered unreasonable in other disciplines. Note, too, that mathematics is purely conceptual. There is no going back and forth, as in the other sciences, between mental ideas and our experience of the world; the mathematician is confined solely to ideas in working out the details of his demonstration. Mathematical ideas may be inspired by reality, of course, and the search for a proof may lead the mathematician to reflect deeply upon the implications of experimental results, but the reasoning itself belongs solely to the conceptual realm.

Mathematics is also an extraordinarily effective science. As everyone knows, most of the major scientific and technological achievements of our time contain some amount, small or large as the case may be, of mathematics. Taken together, these shares measure the power of pure reasoning to exert dominion over matter and the real world—a power so great that some feel reality itself must be, at bottom, an abstract mathematical construction!

For this reason, if mathematics were an art, and if one were to try to identify the art most similar to it, one might well think of design in its many aspects. For in design one encounters the same ambiguity as in mathematics, the same duality—or dialectic—between, on the one hand, harmony, abstraction, and aesthetic appeal, and, on the other hand, the obligation to satisfy a practical purpose. If the design for a table is both elegant and soundly conceived, the table that is made from it will be, as one would hope, handsome, solid, and useful. It is the same in mathematics: a result must be at once original, beautiful, and capable of being applied if its power is to be fully appreciated. We know this from everyday experience, whether we are listening to the weather forecast, or printing out a travel itinerary, or relying on machine translation to read something in another language. All sorts of things that are part of our daily lives would not exist without mathematics. Mathematics is all around us, like tables and chairs, everywhere to be found, in nature no less than in technology; but when it does its job well, we scarcely notice it. In order to send a text message or use an Internet search engine, we do not need to know the underlying mathematical principles, any more than we need to know the principles of electromagnetism, electronics, materials science, and so forth.

And what if mathematics were a literary genre? In that case it would certainly be poetry. This is what Senghor instinctively perceived in contemplating the sybilline language of a series of lecture topics.[4] But I was also made aware of this analogy by readers of my book *Birth of a Theorem,* in which I reproduced verbatim the conversations of professional mathematicians.[5] In literary writing, a poetic quality can arise from the appearance of foreign and unexpected elements. One may find beauty in words whose mysteriousness is the result of hearing them used in an unfamiliar context, for example, where they seem to be out of place. This is a little like what happens when you listen to a song in a foreign language that you do not understand, but in which you sense an occult and melodious force—and where the translation, by dispelling the aura of mystery, will certainly be disappointing! Indeed, Lautréamont experimented not unsuccessfully with the use of mathematical words in poetry in his *Chants de Maldoror.*

Many words that belong to everyday language have been given a special meaning by mathematicians. A ring, for example, is not a band you put on a finger; it is an algebraic structure admitting of two operations, addition and multiplication, having the properties of associativity and commutativity, and incorporating a neutral element. So, too, "spectrum" and "body" have their own definitions in mathematics, along with hundreds of other ordinary words. And when nonspecialists hear these words, which have been adapted by mathematicians to suit very particular purposes, they are apt to discover a kind of poetry in them.[6]

Setting mathematical objects in an unfamiliar context can also be done visually with shapes, as Man Ray did in photographing the collection of three-dimensional mathematical models at the Institut Henri Poincaré in Paris in the 1930s. These objects illustrate the geometric properties of mathematical equations. Ray did not understand the equations at all, but he saw a certain beauty in them; not only did he appreciate the formal elegance of their contours, their distinctive aesthetic quality, but he was fascinated by the fact that they had been made by human hands in order to represent concepts created in human brains. Intuitively he perceived their significance, though it was beyond his understanding. It is much the same when you listen to a foreign language: not only do you hear sounds that have a strange and captivating melody; you know that they have a meaning that some of your fellow human beings are capable of grasping, and this makes them all the more intriguing.

Working from the photographs he had taken, Ray went on to produce paintings of mathematical objects in which one seems to make out faces, masks.[7] These portraits, together with those of many other objects, constitute the series known as *Shakespearean Equations*. In each case the artistic result refers to something other than the original object, something that speaks to all of us, not merely to a small circle of initiates.

Let me conclude these preliminary observations by citing another author, a great poetical novelist and an excellent mathematician by the name of Charles Dodgson, better known as Lewis Carroll. Many people were surprised to learn that he wrote children's books in addition to works of logic. Austere and conservative by temperament, Dodgson very seldom permitted himself to give any hint in his public life of the richness of his imagination. Nevertheless I am certain that, to his way of thinking, everything was connected: the Alice tales are filled with mathematical concepts, logical puzzles, nonsense poems, portmanteau words, and neologisms constructed from carefully devised rules.[8] His logical work is likewise remarkable for its inventiveness and, in places, something very much like magic. Owing to his singular artistry he was able to give mathematics an unmistakably poetic cast.

# 2

## Constraints and Creativity

Not the least of the things that link poetry and mathematics is the importance of constraints. Constraints are indissociable from creativity; indeed, as I have said more than once in my public lectures, they are one of seven conditions for the emergence of new ideas.[9]

In mathematics, constraints are everywhere to be found. In one sense, mathematics is the science of rules and what can be deduced from them. It prides itself on working with as few rules and postulates as possible (think of the centuries-long struggle to do without Euclid's fifth axiom—the parallel postulate—until finally it was shown by Gauss, Lobachevsky, and Bolyai to be independent of the first four axioms); but it requires above all perfectly logical reasoning, and this in itself is an extraordinary constraint. The greatness of mathematics is a consequence of its inventiveness: it has managed, on the basis of a very small number of elementary propositions and despite a great many restrictions, to arrive at a very large number of true statements.

In poetry, whether one considers ancient prosody or the conventions of modern rhyming verse, rules play an extremely important role. What is more, by devising new constraints it becomes possible to invent new literary genres. Here one thinks of the celebrated experiments of OuLiPo,[10] a group that brought together poets, novelists, and mathematicians more than half a century ago and that still exists today. Raymond Queneau, the group's co-founder, demonstrated that by using combinatorics to arrange verses in different forms one could create at least one hundred thousand billion poems.[11] And what happens when one applies arbitrary mathematical rules to a text? The results may be both surprising and beautiful. Take for example the S + 7 rule: whenever you come across a word that interests you, be it a noun, adjective, or a verb, look it up in a dictionary and then replace it by the seventh word that comes after it. Seems simple enough, doesn't it? Applying it to the opening lines of Jean de la Fontaine's "La Cigale et la Fourmi" [The Grasshopper and the Ant, 1686],

> La cigale, ayant chanté
> Tout l'été,
> Se trouva fort dépourvue
> Quand la bise fut venue.
> Pas un seul morceau de mouche ou de vermisseau.
> Elle alla crier famine
> Chez la fourmi sa voisine...

one obtains the following:

> La cimaise ayant chaponné
> Tout l'éternueur
> Se tuba fort dépurative
> Quand la bixacée fut verdie:
> Pas un sexué pétrographique morio

De moufette ou de verrat.
Elle allea crocha frange
Chez la fraction sa volcanique...[12]

Curiously, one can still make out the words of La Fontaine's famous fable even though they have been deformed by substitutions that make a glorious absurdity of it.[13] Queneau's version is both beautiful and amusing; and knowing that it is the consequence of applying a mathematical rule only makes it the more beguiling. That so simple a rule can turn the world upside down and inside out is a delightful thing, in poetry no less than in mathematics.

This interest in constraints is carried on today in a different way in the work of the experimental comic book artist Étienne Lécroart (a member of OuBaPo,[14] an offshoot of OuLiPo), whose illustrations appear at the head of each section of my essay. In one of Lécroart's books, folding the pages in a certain manner joins together parts of successive panels in such a way that a new story emerges with an entirely different meaning. Some of his books are palindromic, so that they can be read from beginning to end, or vice versa, with the same images and the same speech bubbles; some obey more complicated rules. Amazing stuff!

Another way of introducing mathematics in a poetical context is to juxtapose incongruous concepts. Boris Vian, in one of his Collège de 'Pataphysique writings, had the apparently mad idea of trying to work out the numerical calculation of God.[15] He proposed different methods. Depending on the case, they might yield $1 + x$, or 0, or something more elaborate—the result of applying various rules that, though they are perfectly arbitrary, give an impression of logical coherence thanks to a procedure akin to the linking verses of the French children's song "Trois petits chats":

Marabout,
Marabout,
Marabout, -bout, -bout,

Bout de ficelle
Bout de ficelle,
Bout de ficelle, -celle, -celle...

Boris Vian loved mathematics. Like the poet and composer Léo Ferré, he sometimes slipped bits of it into his songs. It is common in France, Vian said, to boast of knowing nothing of mathematics. He had only contempt for this attitude. "I think that someone who understands nothing of mathematics is an incorrigible idiot. I don't see any point in priding oneself on understanding nothing of mathematics and being an idiot." Vian was not a man to mince words, to the displeasure of many. But mathematicians—who have heard people say a thousand times, and with more than a touch of pride, "Oh, I never understood a thing about math!"—understand his annoyance perfectly.

# 3

## Inspiration

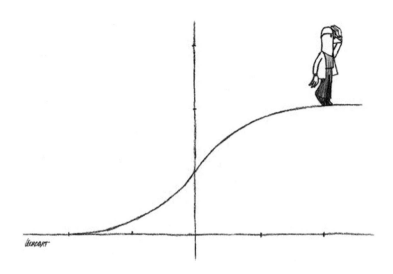

Another common feature of mathematics and poetry is the importance of inspiration. What is more, a mathematical concept can inspire a poetical work of art (I am not aware of an example of a poetical work inspiring a mathematical concept, however). One thinks particularly of M. C. Escher, whose pictorial experiments explored a great many mathematical notions: recursivity, for example, which is the property of applying to oneself, as in his famous picture of two hands drawing each other; or non-Euclidean geometries, which inspired him to create astonishing patterns of interlocking paving stones; or relativity, as in his engraving of the same name, where one abstract concept leads on to another as in a dream. These works do not popularize a mathematical result in order to make its meaning more easily understood, nor do they necessarily use a mathematical rule in order to produce a mathematical result. They are a consequence instead of drawing inspiration from mathematical ideas in order to create works that have an artistic existence of their own.

I have occasionally tried to do the like of this myself, for example in attempting to illustrate the apparent absurdity of the Scheffer–Shnirelman theorem.[16] This paradox—the most astonishing result in fluid mechanics—states that a non-viscous, incompressible fluid can suddenly become violently agitated without any external force having been applied to it.[17] The mathematical formulation is rather more forbidding (incompressible Euler equations admit a non-null distributional solution with compact support in space-time). Lécroart's humorous version of it at the start of this section gives visual expression to an idea whose meaning otherwise would remain locked away in its own private world.

# 4

## Making Connections

The links between mathematics and poetry that I have mentioned so far involve something external to mathematics: the intervention of a reader or an author, the application of mathematical rules to literary texts, the representation of a mathematical idea in physical terms, and so on.

One may therefore wonder whether mathematics has an intrinsically poetic nature, independent of an artist's vision of it. We tend to think of mathematicians as working alone, or at least as part of an isolated community, cut off from the world in which most people live. But even if they do not usually seek to communicate with a wider audience, can they nonetheless be said to approach their work in a poetic way? The reader will not be surprised to learn that, as far as I am concerned, the answer is a resounding yes.

First, because in mathematics one is forever searching for connections, analogies, comparisons—a classic way of proceeding in poetry. I tried to give this idea a nested structure in my book *Birth of a Theorem*. There a mathematician (yours truly) listens to a song by the French singer Gribouille, based on a poem by Jean-Marie Huard, and then takes stock of his attempts over the past twelve years to discover unsuspected connections between different areas of mathematics—which makes him think once more of the song.[18] In his work he establishes connections between very different objects; and so, listening to Gribouille, he perceives a connection between his work and the song, because in the song a connection is made between a sailor and a rose, which one should have thought had nothing to do with each other.

Establishing connections between things that appear at first sight to be entirely unrelated has a long and glorious history in mathematics. Gauss, Galois, Riemann, Poincaré, Bachelier, Noether, Serre, Grothendieck, Gromov, Wiles, Langlands, Thurston, Lieb and Thirring, Lawler, Schramm, and Werner—these are only a few of the celebrated figures who owe their renown to the fact that they discovered correspondences whose existence no one had suspected.

In my own research, I have constantly been inspired to search for connections, between a problem in geometry and another in statistical physics, for example. I was fortunate enough to be part of a group of mathematicians who discovered a close relationship between non-Euclidean geometry (Ricci curvature), optimization (optimal transport), and statistical physics (entropy). Here, as in so many cases, one is looking for a way to unite concepts developed by different persons having different aims and working at different times with different techniques and different theories, often formulated in different ways. This approach is responsible for many advances in mathematics and for many famous results; it illuminates all the constituent parts of a problem, and sometimes makes it possible to solve the problem in a spectacular fashion. "Mathematics," as Poincaré said, "is the art of giving the same name to different things."[19] Why should this art be the province of mathematics rather than any of the other sciences? Perhaps because mathematicians manipulate abstract ideas, and because conceptual abstraction is often inherent in physical phenomena of various kinds.

One of the most striking results in all of mathematics, and perhaps in all the sciences, is the central limit theorem (or, as Sir Francis Galton called it, the law of frequency of error). The theorem says that the result of adding independent random errors together (an appropriate level of measurement having been selected beforehand) has a universal profile, a bell-shaped curve representing the high probability of small errors and the very low probability of large errors. The same curve is observed to obtain in political opinion polls, the fluctuation of water levels, the movements of particles subjected to constant agitation, and variations in human height distributions—phenomena that no one would have thought were related. An abstract idea, in other words, may assume a variety of forms in the physical world. The same impulse lies at the heart of poetry. Poets likewise establish a relation between different things, between an object and a moment of everyday life, for example, by means of images, allegories, and analogies of different sorts.

# 5

## A Portable Universe

Mathematics and poetry can be compared for another reason: each aspires to recreate a universe—a portable universe, a universe that we can carry around in our heads. The mathematician transforms some aspect of the physical world into a few equations that he can hold in mind before setting to work on paper; in the same way, a poet recreates a world within the restricted space of a few stanzas, permitting his readers to make it their own.

Some aspects of reality are so inaccessible that they too can be likened to fables or abstractions. Mathematics gives us a way of thinking about them more precisely in the form of formulas. Think of planets such as Jupiter, or of stars millions of light-years away that we will never visit. Nevertheless we are capable, by a process of abstraction, of representing the forces that govern their movement through Newton's law of universal gravitation. In this way we can predict their future course for aeons to come. Indeed, one might go so far as to speak of recreating a universe in a space so conceptually restricted that it can, in a sense, be mastered. In Einstein's famous phrase, the most incomprehensible thing about the universe is that it is comprehensible.[20] Mathematical re-creation, by means of formulas and equations, is related to poetry in its root sense: etymologically, poetry—from the Greek *poiēsis*—means creation.

I like to begin some of my lectures by alluding to a famous poem, "The Lady of Shalott" by Tennyson. The poet recounts an Arthurian legend in which an unfortunate young noble woman, who has been unjustly placed under a curse, finds herself incapable of looking at the world directly with the naked eye; she can only observe it through the reflection of a mirror. One day she sees a fabulously handsome knight, Sir Lancelot, pass by the tower in which she is imprisoned. Instantly smitten, and unable to prevent herself from looking at him directly, she is condemned to die. This tragic ballad has been interpreted in many ways. What was Tennyson really trying to say? I like to imagine (after all, when it comes to poetic exegesis, one is free to imagine what one likes in the absence of any explicit statement by the author) that his tale is an allegory of mathematics—personified by a young woman! Mathematicians, unable to apprehend the world directly by means of experiments, as physicists do, are condemned to study it through its mathematical reflection: equations.

# 6

## The Form of Words

Poetry places the highest value on words and the form of words, for their evocative power, for the impressions they are capable of arousing in the mind of the reader. If there is a science in which language has comparable or even greater importance, it is unquestionably mathematics. Mathematics is itself a language; indeed, it is the language par excellence of the exact sciences.[21] Physicists, though they seldom display the same degree of rigor as mathematicians, rely on its formalisms in order to express their discoveries and make them widely known.

What is more, mathematics has become a universal language, used everywhere in the world today, one of the very few universal languages that currently exist. In a sense it is more universal even than music, for while all peoples cherish music, musical conventions vary considerably from one culture to another, whereas mathematical conventions are everywhere the same, or very nearly so. Mathematics is a universal language in which one knows exactly what has been stated, and in which symbols have a special significance, not only for purposes of verification, of course, but also for communicating very well-defined ideas.[22] If I say to you "epsilon," for example, you think of the Greek letter; if I tell you that a number is called epsilon, at once you say to yourself: it must be small. All mathematicians know that it designates a very small number, which means that it is going to figure in an argument in which the possibility of choosing an arbitrarily small value will be crucial. Every domain of mathematics has similar conventions: $f(x)$ stands for a function (evaluated for a variable $x$), $m$ stands for the slope or gradient of a line, and so on. Mathematical conventions, no less than poetical conventions, transmit meaning. Poetical conventions make it possible to convey impressions and arouse expectations; and by artfully manipulating semantic fields, they create imaginative contexts for them.

# 7

# Visionaries

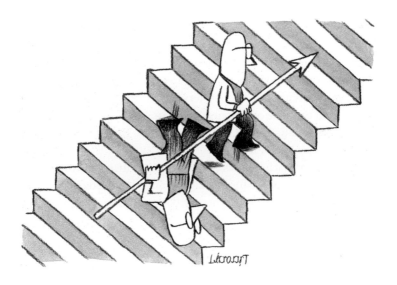

Mathematics, above all else, is a creative science. Compared with a good mathematician, an excellent mathematician is someone who creates, comprehends, and reorders—someone who sees things in a new way, just as an excellent poet sees something extraordinary in an ordinary object and finds a new way to communicate it using words and images.

To take only one example among a great many (don't worry if you don't understand what it involves), the astounding proof by the Romanian mathematician Dan-Virgil Voiculescu of the

Shannon-Stam inequality (a famous relation in information theory) using the Brunn–Minkowski inequality (a classic theorem in geometric analysis)[23] cast an altogether unexpected light on both results, forcing specialists to think long and hard about the meaning of things they thought they understood very well. Grothendieck's amazing proof of the Riemann–Roch theorem[24] utterly changed the way mathematicians thought of this statement. Genius, it has been said, is the discovery of something that no one had even imagined, something that is suddenly brought to light for the first time.

Sometimes it is the conclusion the mathematician has reached that is a source of wonder, sometimes it is the tool he has used. Gromov dumbfounded geometers by showing the powerful use that could be made of the most elementary tools, such as the triangular inequality. What astonished them, in other words, was his incomparable creativity.

Success in mathematics depends quite obviously on having an enormous appetite for work. But one also needs inspiration. Without inspiration, no mathematician will go very far. The same is true of poets. "No one can be a mathematician who does not have the soul of a poet," said the great Russian mathematician Sofia Kovalevskaya.[25] Inspiration can come from any number of things, an observation, a new concept, a calculation. Sometimes reading a new article fills mathematicians with a sort of giddiness, a singular feeling compounded of admiration and awe in the presence of a stunning insight. "How in the world did he think of that!?!"

Consider John Nash, a towering figure in the modern history of both mathematics and economics. All statisticians are familiar with entropy, but Nash interpreted it in such a way that unexpected insights into regularity could be obtained from the study of dissipative equations. Grigory Perelman later found a way to use entropy as a delicate analytic control, one of the keys to his solving the Poincaré Conjecture. Both Nash and Perelman were able to look at entropy in a new and different light. The greatest mathematicians of recent decades have all shared this ability, which turns high-level mathematical research into a brilliant detective story, with breathtaking twists and turns on every other page.

Remarkably, it is sometimes the beautiful and unexpected character of a mathematician's vision that helps to convince others of its soundness. Here again one thinks of Poincaré's conjecture, a landmark of topology that took almost a hundred years to prove, stimulating research throughout the twentieth century and beyond.[26] In the mid-1970s a farseeing young American geometer, William Thurston, came upon the scene. Thurston did not succeed in proving the conjecture, but he did show that if it were true, it would form part of an extraordinary mathematical landscape with all sorts of features no one had ever dreamed possible—a little as though you had come across a new animal species and a taxonomist said to you, "This is but one species among many, part of a whole new genus, and here is what the genus looks like. I say that other species will be discovered, this one, that one, and that one." If a vision is so beautiful and so harmonious that it takes your breath away, you are bound to fall under its spell—it is so beautiful, you say to yourself, it simply must be true! This is what happened with Thurston. Until he came along, mathematicians were not really sure whether they ought to bother trying to prove Poincaré's conjecture. Thurston showed that it fit into a vast and previously unimagined landscape, a landscape so magnificent that everyone began to believe it had to exist. By reinterpreting Poincaré's conjecture he had managed to bring forth a new world for all to see. In this sense, Thurston was a poet.

Moreover, the word *poetry* has sometimes been employed by scientists themselves as a sign of admiration. Lord Kelvin, the greatest physicist of his time, called Fourier's theory a "mathematical poem." The parallel between mathematical creativity and poetic creativity can be extended still further if we take into account the passion that accompanies creation, the anxiety one feels before the blank page, the ebb and flow of inspiration, and the role played by intuition. Poincaré said that the most important thing parents can do is to inculcate in their children a sense of wonder at the marvels of nature. Certainly a mathematician's most precious possessions are a talent for being fascinated by mathematical problems and for being enchanted by mathematical beauty.

Nevertheless I would rather conclude this chapter by emphasizing the notion of style, something that is prized by artists, and quite obviously by poets, but also by mathematicians. Mathematicians have different styles—in the way they approach problems, in the way they state them, in the way they solve them. Grothendieck wrote at some length about this, and magnificently, in his *Récoltes et semailles*,[27] comparing his style of solving mathematical problems to immersing a nut in an emollient liquid so that the shell splits open by itself, effortlessly. Poincaré's untidiness, Thom's sibylline concision, Bourgain's technical extravagance, Doob's clarity—all these things have long been remarked upon by mathematicians, along with many other aspects of style. The luminous and informal manner of the young Hörmander, for example, had little in common with the austere manner of the older Hörmander. In my own career as a mathematician I have borrowed from the style of various authors whom I admire, rather like an artist who finds inspiration in the work of one artist and then of another, and who creates his own style by building upon them, partly in reaction against his predecessors while at the same time following their lead. Like Hörmander and so many others, my style has evolved over the years—without my being able to do much about it![28]

# 8

## Poincaré and the Omnibus

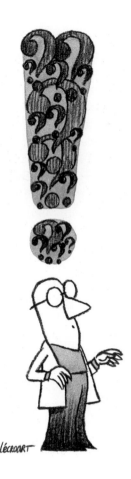

Where does inspiration come from? Not from applying a rule. Often it comes at a decisive moment, after long and unremitting effort. Here again I do not think there is any very great difference between mathematicians and poets. In my book *Birth of a Theorem* I mentioned a few instances of sudden inspiration, some seeming to come out of nowhere, some triggered by a manifest sequence of events, or else by a certain atmosphere or mood, or by an apparently insignificant action.

On this subject I defer to Henri Poincaré and his famous essay on mathematical discovery, which forms the third chapter of the first part of *Science and Method* (1908) and is reprinted here in the Appendix, following my own essay.[29] Poincaré describes the role of unconscious mechanisms and analogies in describing a mathematical situation, the difficulties encountered in establishing a relation between distinct elements, the importance of bringing a new perspective to bear on an object, and the role of inspiration itself, which may be drawn not only from prior discoveries, but also from one's surroundings, from reading a poem or listening to a piece of music, or from something else altogether.[30]

Poincaré's essay is one of the best accounts we have of what goes on in the mind of a great mathematician when he is grappling with an intractable problem. Note that Poincaré deliberately inserts scientific jargon in a book intended for a nonspecialist audience. Normally, one should have thought that the author must not allow the reader's attention to be distracted by extraneous details. If the reader is busy trying to work out what Theta-Fuchsian functions are, for example, he is liable to overlook apparently small details—exactly the things that the author wants the reader to concentrate on, his drinking black coffee one evening, getting on the omnibus, and so on. Poincaré circumvents the difficulty, as I did just a moment ago, by assuring his readers at the outset that it does not matter whether they understand the technical details.

Note also the alternation, well known in scientific research, between periods of systematic and conscious exploration, on the one hand, and flashes of insight, on the other, which tell you what direction to take. These unconscious mechanisms can be triggered by anything at all. High school students would do well to remember this, as they may find it very useful. If you're having trouble with a difficult math assignment that has to be turned in tomorrow morning, and yet you simply have to go out with your friends this evening, you can always tell your parents that you are only following Poincaré's example—searching for inspiration!

# 9

## Ping-Pong

Observe, too, the role of analogy in Poincaré's essay. One can detect a poetic element in the way in which he describes his work by means of images borrowed from military operations: laying siege to a problem, which is likened to a system of fortifications, then capturing its outworks one by one until, with the fall of the last, the central fortress is finally captured. On this view, the mathematician is a general who devises strategy and marshals his forces. When you are writing about science for a popular audience, it is often wise to use images that are well known to everyone.

In a recent magazine article, I drew upon an analogy with a sport that I loved to play when I was young, ping-pong (sorry, table tennis). I tried to evoke the interactions that are so much a part of what scientists do, not only the daily exchanges among close collaborators, but also the ones that occur when one presents one's results, whether by giving talks to small groups of colleagues or by taking part in conferences and seminars. All these things can be seen as forming a special sort of competition:

> My paddle, one side red, the other black, decorated with the butter-fly dear to Ping-Pong players, was my proud and faithful ally during thousands of hard-fought matches. Twenty years ago, no longer having the time or the desire to take on new adversaries, I hung up my racket. But the ricocheting serves, the plunging topspin forehands, the slashing returns, the tense rallies, the non-chalant flicks of the wrist, the smashes and the saves—all these shots played themselves over and over in my brain long after I retired, their trajectories picturesquely bent by the Magnus effect.
>
> In the meantime I had devoted myself in earnest to mastering the sinuous trajectories of theories and theorems—so many balls that scientists and mathematicians hit back and forth to one another, sometimes sharply, sometimes gently, sometimes with great violence, in a vast and very ancient collective match of ping-pong.[31]

In the case of mathematics, the match takes place within differ-ent branches of the discipline, but also across all of them—the vast collective match I referred to. But note, too, that I smuggled a technical term into this passage: the Magnus effect. In physics, the Magnus effect is observed in spinning objects that follow a deflected trajectory. It is owing to this effect that a ball hit with top spin will swerve downward, because it turns rapidly in the direction of travel. Even if you hit the ball very hard, then, it is apt to stay within the lines of a tennis court or the edges of a ping-pong table. A ball sliced from beneath, a softer shot, tends to float and is less likely to land on the table. If you hit it just right, your opponent will have trouble returning it; but it is harder to control and to keep in play. The Magnus effect is fundamental in golf, tennis, and table tennis, where it changes the entire complexion of the game. The same is true of soccer, though to a lesser degree; one finds it in the swerving trajectories of free kicks, for example. An enormous proportion of soccer players, maybe as high as 99.9 per cent, learn to master this effect without knowing what it involves.

# 10

## Ode to Imperfection

The best way to illustrate a scientific concept or theme is to tell a story. There are really only two things that capture the attention of people everywhere: games and stories.

In talking about science, it is often helpful to retrace a historical path of development. With regard to the age of the Earth, for example, one may relate how our understanding evolved from ancient times by describing the increasingly precise theories that were advanced and the great controversies that grew up around them; in the nineteenth century, they set Darwin against Kelvin, proving that even great minds can be completely mistaken, with the result that an opportunity for making considerable progress in geology was missed. Unraveling a narrative thread of this sort, as I have done many times in my lectures, makes it possible to show the connections between concepts.[32]

Improvising on a set topic can also be a rewarding exercise. In July 2012 I took part in La Milanesiana, a festival held every year in Milan devoted to promoting a dialogue among philosophy, literature, cinema, the performing arts, and the sciences.[33] Each edition of the festival has a particular theme, and scientists are invited to give talks in a poetic vein. It is an occasion for them to give non-scientists a sense of what their work involves by arousing certain impressions and emotions in their listeners. The theme in 2012 was imperfection. Here, then, by way of conclusion, I leave you with the text of my talk, in which you will find once more some of the subjects we have discussed in the preceding pages: creation, constraints, beauty, and the great Henri Poincaré.

## Ode to Imperfection

Henri Poincaré—the foremost mathematician not only in France but in the whole world, as it used to be said in his homeland and everywhere else—passed away one hundred years ago.

A mild-mannered man from a distinguished family, heavyset, blind as a bat, who by the sheer force of his intellect managed to raise questions that human beings will ponder for centuries to come, Poincaré was not only a great mathematician; he was also a great physicist, a great astronomer, a great engineer, a great philosopher—in short, a universal mind, consulted in his last years as an oracle on all subjects. His memory lives on as a symbol of the force and unity of human thought, fragile and precious, about which he wrote at length with great insight and eloquence. "Thought," he says in the concluding lines of *The Value of*

*Science*, "is only a gleam in the midst of a long night. But it is this gleam which is everything."[34]

Poincaré was interested in everything, learned all there was to know, and revolutionized thinking in mathematics and physics. He saw everything on a grand scale. No wonder, then, that when he was wrong, his errors were on a grand scale as well. After all, only the dead make no errors; and Poincaré was not a man who was willing to confine himself to timid expressions of opinion that do not have the virtue even of being false.

He committed his most famous error, which will long stand out in the annals of science, while studying the three-body problem. Solutions to the problem of interacting celestial bodies had been known for two bodies since Newton, but not yet for three bodies or more. Take only the case of two massive bodies, the Earth and the Sun, for example. Ignore the rest of the solar system, and calculate their movements with the aid of Newton's equations. The solution is readily found: the Earth traces a marvelous ellipse around the Sun, a simple and elegant trajectory already known to Greek mathematicians more than two millennia ago, before it was understood that the Earth turns on its axis; and then rediscovered by the German astronomer Johannes Kepler in the early seventeenth century, before Newton elucidated the phenomenon of gravitational attraction.

With two bodies, then, we have a lovely, permanently stable ellipse, endlessly repeating itself. But what happens when we take into account the movements of the other planets? After all, if the Earth is irresistibly attracted by the Sun, it is also influenced by Mars, Jupiter, and more distant bodies as well. To be sure, the combined influence of these bodies is not very great by comparison with the attraction of the Sun, but would it nonetheless be enough to deflect the Earth from its course? Would the Earth continue forever to go around the Sun undisturbed, or might it one day collide with another planet? Once the influence of a third body was considered, nothing was clear any longer, no one had any idea what might happen—and in a solar system with nine or ten planets (as it was then supposed), the situation was still more intractable. Let us confine our attention to three bodies and look at what the equations tell us. Do we find stability or instability?

Poincaré took up a slightly simplified version of the three-body problem when he was in his early thirties, in the hope of winning a prize announced by King Oscar II of Sweden in 1885. The problem was bound to fascinate a talented young mathematician who loved nothing more than to contemplate the fundamental laws of nature—and all the more as it presented him with an opportunity to outdo himself! Three years later, the jury had no trouble guessing the identity of the anonymous author of a manuscript brimming with new ideas and neologisms that demonstrated stability in so elegant a manner. Poincaré won the prize hands down.

Nevertheless his submission was not perfect, far from it. Vagueness, imprecision, and ambiguity abounded in Poincaré's demonstration. Nothing surprising about that. Everyone knew that, for all his undoubted brilliance, Poincaré was not a model of clarity. Elliptical phrasing, unjustified assertions, learned digressions interrupting the rhythm of the argument—all these things were familiar to Poincaré's readers. Verification was by no means easy, and the long list of queries drawn up by Lars Edvard Phragmén, the talented young assistant editor at *Acta Mathematica* responsible for preparing Poincaré's manuscript for publication, raised no eyebrows.

Poincaré corrected as much as he could, convincing himself that finally everything was in good order, and that the edifice he had constructed so carefully would withstand any further assault. And yet one of the cracks in the façade that Phragmén had spotted began to trouble Poincaré more and more each day. Eventually he had to admit the obvious: he had gotten everything wrong!

A tiny fissure had become a gaping hole. The entire structure collapsed of its own weight.

But Poincaré had already been awarded the prize, in January 1889. The honors and money were his, the article had been published, his achievement was celebrated on all sides. The pressure weighing on him was phenomenal. What could be done?

To begin with, every effort had to be made to prevent the fatally contaminated proof from being widely noticed. Happily, the Internet did not yet exist! In the end the journal's editor-in-chief managed to get back every copy of the issue containing Poincaré's article. The entire print run was pulped, with Poincaré having to bear the full cost. Nevertheless he emerged from the affair with his reputation unblemished. At once he set about trying to remedy the defects of his earlier work.

Amazingly, Poincaré succeeded in fixing everything, only now on a larger scale, of course, since his conclusion had completely changed. In putting his finger on the source of his original error he was able to discover how instability could emerge in a clockwork system of celestial motions governed by implacably precise equations—equations more exact than the most reliable Swiss watches, but applying to a system so sensitive to initial conditions that its actual path could be modified by the fluttering of a butterfly's wings, as it would later be said.

With the aid of another French mathematician, Jacques Hadamard, whose work helped to consolidate Poincaré's new results, forcing Keplerian perfection to give way to a sublime imperfection, vast new vistas were opened up. Just as Christopher Columbus inadvertently stumbled upon America, Poincaré discovered a new scientific continent, an imperfect and chaotic world whose laws, even if they are, at bottom, deterministic, lead to unpredictable behaviors that can only be apprehended in statistical terms. But chance, Poincaré hastened to point out, must be something more than what we are accustomed to call ignorance. "You ask me to predict the phenomena that will be produced. If I had the misfortune to know the laws of these phenomena, I could not succeed except by inextricable calculations, and I should have to give up the attempt to answer you; but since I am fortunate enough to be ignorant of them, I will give you an answer at once. And, what is more extraordinary still, my answer will be right."[35]

A great discovery, in truth—and all the more beautiful, I should say, for having been the offspring of a grave error. An error that, in hindsight, no longer seems to be so grave as it once did, and one that left a lovely birthmark; an imperfection that is part of the charm of the theory of deterministic chaos, just as the webbed hand of the young woman in Kafka's *Trial* is part of her beauty.

And yet, in spite of his dramatic reversal, Poincaré did not go so far as to call into question Newton's fundamental laws. For the most part, at least, they were still intact.

Ten years after Poincaré's magnificent mistake, as the twentieth century was getting underway, scientists could congratulate themselves on having identified all the laws of physics. For the first time, humanity had at its disposal coherent theories explaining everything: mechanics, astronomy, electromagnetism, fluids, waves, and much else besides. True, there were one or two unsettled points. The Michelson–Morley experiment, black-body radiation— small rough spots on the surface of a gigantic diamond. The diamond would have to be polished.

Alas! These minor imperfections, once brought to light, swiftly assumed enormous proportions that exceeded the ability of diamond polishers to polish. Controversy grew over the so-called ultraviolet catastrophe—an upheaval that soon produced not one, or two, but three revolutions in theoretical physics: radioactive transmutation of the elements, relativity, and quantum mechanics. It took three decades to explore these new horizons, in which light, energy, and matter, having become improbable synonyms, now shone with an unprecedented intensity.

Nietzsche said that one must still have chaos in oneself to be able to give birth to a dancing star.[36] Poincaré had shown that Newton's physics still contained chaos: a deterministic system could nonetheless give birth to unpredictability. Fundamental physics at the beginning of the twentieth century, despite its mask of perfection, still contained enough chaos to give birth to not fewer than three dancing stars.

Hold on a moment. A dancing star? A lovely phrase, to be sure, but doesn't it call to mind something that sparkles with perfection, something beautiful and sublime, swaying to the music of the spheres? Nonsense! The stars are a vast confusion, they embody the reign of instability. Have you ever stopped to consider that whereas gases are organized in a homogeneous fashion and spread everywhere harmoniously and uniformly, stars are concentrated in irregular clusters, separated by huge expanses of empty space? Stars organize themselves in galaxies, the galaxies in clusters of galaxies, and the clusters in superclusters. Far from being harmonious and regular, the distribution of stars is lumpy, perhaps actually fractal.

We may not know who composed this strange stellar ballet, but we do know the orchestra conductor: Newton's equation, in the form of its statistical version, Vlasov's equation. It is through these equations that the properties of stars can be identified. Mathematical analysis furnishes us with the key to unlocking the secrets of their irregular behavior: Jeans instability, which prevents homogeneous matter from being stable at large wavelengths—an instability now raised to the status of a law, a mathematical *clinamen* (Lucretius's name for the unpredictable swerve of atoms) from which the entire structure of the Universe was generated.

Music, for its part, is no more perfect than the movement of the stars. From the time of Pythagoras, and perhaps much earlier, music has been a mathematical art based on relations among frequencies. At 440 pulses per second you hear the note A, and if you double the frequency to 880 pulses you will hear an A one octave higher; each time you double it, you go up an octave. And if you triple the frequency to 1320 pulses you will go up to the fifth in the next higher octave, which is to say E. With only the factors of two and three, in other words, you can move from octave to octave and from fifth to fifth.

But no matter how hard you try, you will find that it is impossible to create a perfect scale in this manner. For a series of doublings will never yield a series of triplings; a power of two will never equal a power of three. At some point it will become necessary to cheat in constructing the scale, to decide, as if by decree, on a constitutive imperfection—by introducing minute differences, such as the Pythagorean comma, a tiny interval that breaks the natural symmetry of a system of tuning, or by disrupting the exact arrangement of frequencies and systematically introducing irrationality, in the mathematical sense of the term. Our music is inherently imperfect, in other words—and yet so rich, so pleasing, so full of possibilities!

As human beings, of course, we are well acquainted with imperfection. We are steeped in it; indeed, it might be said that we are its children, for we owe everything to imperfection. It is owing to the imperfection of reproduction that species were able to evolve; hundreds of millions of genetic mutations have occurred since the emergence of bacteria and made us what we are, having been selected across all human populations, from transcription error to transmission error. As the French singer Mama Béa Tékielski put it, "We are the result of a false equation." Lucky us! Imperfection, statutory and salutary, is our strength; were we all perfect, we would be condemned to extinction. Genetic variability is our greatest advantage in confronting the constantly changing dangers of the biological world. What is more, it produces marvelous admixtures and amalgamations!

There is imperfection in everything we do and make. Not least in our languages, whose fabulous diversity is the consequence of innumerable errors of textual reconstruction, mistakes of spelling and grammar, distortions of speech patterns and mispronunciations of all kinds; of unsettled dialects giving way to tenacious accents (thus bad Latin eventually became good Italian, for example). How many hundreds of thousands of errors enshrined in correct usage have combined to erect our monumental tower of Babel!

Imperfection lurks in our computer programs, which grow more complex by the day. No one will ever be able to eradicate all the bugs that infest them. It lurks also in our technological gadgets, whose congenital errors of design live on and on. No amount of progress will be able to fix them. Think of our word processing machines, saddled, perhaps permanently, with an absurdly inefficient keyboard.

But isn't human thought perfect, this gleam, this flash of lightning in the dark night? Please! Nothing is more confused! It was only with the greatest effort that human beings managed to invent mathematical reasoning, perfect in its form and its logic; but perfection is not its primordial condition. Poincaré saw this clearly, looking back upon some of his most magnificent discoveries, in which more or less long periods of conscious reflection led on to spontaneous and inexplicable associations of ideas—an unpredictable sequence of events that mirrors the very chaos implied by his physical theories.

"Error led him by chance to the truth," Voltaire said of Kepler. The same might be said of Poincaré, or Wiles, or many others, without in any way underestimating the formidable combination of effort and imagination that led them into error in the first place. In Kepler's case, the combination was complicated by absurd mystical reveries. "There are such minds as must rely on geometry, and that fall down when they wish to walk alone,"

Voltaire went on to say, by way of excusing Kepler.[37] But is it not the case that mathematicians must also rely on the irrational, on intuition, no matter how strange their counsels may seem?

Even the best mathematicians make mistakes, as we have seen in the case of Poincaré. Sometimes they fall into two errors at once that have the good taste to cancel each other out, as in the case of Galileo describing the trajectory of a canon ball. Sometimes they run up against three errors that are mutually reinforcing, as in the case of Lord Kelvin calculating the Earth's age. Examples and counterexamples of this type can be endlessly multiplied. They show that error is not simply an obstacle we encounter in making our way forward; it is also a source of progress, part of what enables us to advance. There is nothing to be regretted in the least about this. In the domain of human thought, as in that of language and biology, we can only rejoice that error is possible, for it is error that gives rise to the unexpected—and sometimes the sublime!

Consider the role of error in the career of John Nash, another legendary figure of constructive illumination, who proved three theorems in scarcely more than five years that revolutionized the field of analysis, before being awarded the Nobel Prize in economics for his early work on equilibria.[38] When he set about proving his two embedding theorems,[39] in response to a challenge issued by an MIT colleague who was exasperated by his arrogance, he was aware of attempting something very difficult and very important. After a chaotic gestation, he proudly submitted the balance of his results, on smooth embeddings, for peer review to the *Annals of Mathematics* in the fall of 1954. Whereas the first proof concerning unsmooth embeddings, published earlier that year, was clear and briefly stated, however, the second was a meandering argument of staggering complexity, an opaque jumble of ideas whose main themes could barely be made out. It was only through the generous cooperation and heroic efforts of the journal's referee, Herbert Federer, that Nash's manuscript was able to be put into publishable form.[40]

But what an ungainly demonstration! As it happens, there is another way of solving the problem. Thirty years later, the German mathematician Matthias Günther found an elegant and simple solution—a perfect solution!

And yet it was fortunate that Nash had not seen it, this perfect solution. For out of the disordered mass of ideas, some of them false, the whole subsequently reworked and simplified many times, there finally emerged the most powerful technique of nonlinear perturbative analysis yet found, the Nash–Moser theorem, whose importance extends very far beyond the framework of isometric embedding—a universal method that students will continue to be taught for centuries to come.

It is from imperfection, then, that great advances in knowledge are born. A lyric of the Italian singer and poet Fabrizio De André, touched by the grace of inspiration, says this wonderfully well:

> *Dai diamanti non nasce niente*
> *Dai letame nascono i fiori*
> From diamonds come nothing
> From peat come flowers

# Henri Poincaré on Mathematical Discovery

The genesis of mathematical discovery is a problem which must inspire the psychologist with the keenest interest. For this is the process in which the human mind seems to borrow least from the exterior world, in which it acts, or appears to act, only by itself and on itself, so that by studying the process of geometric thought we may hope to arrive at what is most essential in the human mind.

This has long been understood, and a few months ago a review called *l'Enseignement Mathématique*, edited by MM. Laisant and Fehr, instituted an enquiry into the habits of mind and methods of work of different mathematicians. I had outlined the principal features of this article when the results of the enquiry were published, so that I have hardly been able to make any use of them, and I will content myself with saying that the majority of the evidence confirms my conclusions. I do not say there is unanimity, for on an appeal to universal suffrage we cannot hope to obtain unanimity.

One first fact must astonish us, or rather would astonish us if we were not too much accustomed to it. How does it happen that there are people who do not understand mathematics? If the science invokes only the rules of logic, those accepted by all well-formed minds, if its evidence is founded on principles that are common to all men, and that none but a madman would attempt to deny, how does it happen that there are so many people who are entirely impervious to it?

There is nothing mysterious in the fact that every one is not capable of discovery. That every one should not be able to retain a demonstration he has once learnt is still comprehensible. But what does seem most surprising, when we consider it, is that any one should be unable to understand a mathematical argument at the very moment it is stated to him. And yet those who can only follow the argument with difficulty are in a majority; this is incontestable, and the experience of teachers of secondary education will certainly not contradict me.

And still further, how is error possible in mathematics? A healthy intellect should not be guilty of any error in logic, and yet there are very keen minds which will not make a false step in a short argument such as those we have to make in the ordinary actions of life, which yet are incapable of following or repeating without error the demonstrations of mathematics which are longer, but which are, after all, only accumulations of short arguments exactly analogous to those they make so easily. Is it necessary to add that mathematicians themselves are not infallible?

The answer appears to me obvious. Imagine a long series of syllogisms in which the conclusions of those that precede form the premises of those that follow. We shall be capable of grasping each of the syllogisms, and it is not in the passage from premises to conclusion that we are in danger of going astray. But between the moment when we meet a proposition for the first time as the conclusion of one syllogism, and the moment when we find it once more as the premise of another syllogism, much time will some-times have elapsed, and we shall have unfolded many links of the chain; accordingly it may well happen that we shall have forgotten it, or, what is more serious, forgotten its meaning. So we may chance to replace it by a somewhat different proposition, or to preserve the same statement but give it a slightly different meaning, and thus we are in danger of falling into error.

A mathematician must often use a rule, and, naturally, he begins by demonstrating the rule. At the moment the demonstration is quite fresh in his memory he understands perfectly its meaning and significance, and he is in no danger of changing it. But later on he commits it to memory, and only applies it in a mechanical way, and then, if his memory fails him, he may apply it wrongly. It is thus, to take a simple and almost vulgar example, that we sometimes make mistakes in calculation, because we have forgotten our multiplication table.

On this view special aptitude for mathematics would be due to nothing but a very certain memory or a tremendous power of attention. It would be a quality analogous to that of the whist player who can remember the cards played, or, to rise a step higher, to that of the chess player who can picture a very great number of combinations and retain them in his memory. Every good mathematician should also be a good chess player and *vice versâ*, and similarly he should be a good numerical calculator. Certainly this sometimes happens, and thus Gauss was at once a geometrician of genius and a very precocious and very certain calculator.

But there are exceptions, or rather I am wrong, for I cannot call them exceptions, otherwise the exceptions would be more numerous than the cases of conformity with the rule. On the contrary, it was Gauss who was an exception. As for myself, I must confess I am absolutely incapable of doing an addition sum without a mistake. Similarly I should be a very bad chess player. I could easily calculate that by playing in a certain way I should be exposed to such and such a danger; I should then review many other moves, which I should reject for other reasons, and I should end by making the move I first examined, having forgotten in the interval the danger I had foreseen.

In a word, my memory is not bad, but it would be insufficient to make me a good chess player. Why, then, does it not fail me in a difficult mathematical argument in which the majority of chess players would be lost? Clearly because it is guided by the general trend of the argument. A mathematical demonstration is not a simple juxtaposition of syllogisms; it consists of syllogisms *placed in a certain order*, and the order in which these elements are placed is much more important than the elements themselves. If I have the feeling, so to speak the intuition, of this order, so that I can perceive the whole of the argument at a glance, I need no longer be afraid of forgetting one of the elements; each of them will place itself naturally in the position prepared for it, without my having to make any effort of memory.

It seems to me, then, as I repeat an argument I have learnt, that I could have discovered it. This is often only an illusion; but even then, even if I am not clever enough to create for myself, I rediscover it myself as I repeat it.

We can understand that this feeling, this intuition of mathematical order, which enables us to guess hidden harmonies and relations, cannot belong to every one. Some have neither this delicate feeling that is difficult to define, nor a power of memory and attention above the common, and so they are absolutely incapable of understanding even the first steps of higher mathematics. This applies to the majority of people. Others have the feeling only in a slight degree, but they are gifted with an uncommon memory and a great capacity for attention. They learn the details one after the other by heart, they can understand mathematics and sometimes apply them, but they are not in a condition to create. Lastly, others possess the special intuition I have spoken of more or less highly developed, and they can not only understand mathematics, even though their memory is in no way extraordinary, but they can become creators, and seek to make discovery with more or less chance of success, according as their intuition is more or less developed.

What, in fact, is mathematical discovery? It does not consist in making new combinations with mathematical entities that are already known. That can be done by any one, and the combinations that could be so formed would be infinite in number, and the greater part of them would be absolutely devoid of interest. Discovery consists precisely in not constructing useless combinations, but in constructing those that are useful, which are an infinitely small minority. Discovery is discernment, selection.

How this selection is to be made I have explained above. Mathematical facts worthy of being studied are those which, by their analogy with other facts, are capable of conducting us to the knowledge of a mathematical law, in the same way that experimental facts conduct us to the knowledge of a physical law. They are those which reveal unsuspected relations between other facts, long since known, but wrongly believed to be unrelated to each other.

Among the combinations we choose, the most fruitful are often those which are formed of elements borrowed from widely separated domains. I do not mean to say that for discovery it is sufficient to bring together objects that are as incongruous as possible. The greater part of the combinations so formed would be entirely fruitless, but some among them, though very rare, are the most fruitful of all.

Discovery, as I have said, is selection. But this is perhaps not quite the right word. It suggests a purchaser who has been shown a large number of samples, and examines them one after the other in order to make his selection. In our case the samples would be so numerous that a whole life would not give sufficient time to examine them. Things do not happen in this way. Unfruitful combinations do not so much as present themselves to the mind of the discoverer. In the field of his consciousness there never appear any but really useful combinations, and some that he rejects, which, however, partake to some extent of the character of useful combinations. Everything happens as if the discoverer were a secondary examiner who had only to interrogate candidates declared eligible after passing a preliminary test.

But what I have said up to now is only what can be observed or inferred by reading the works of geometricians, provided they are read with some reflection.

It is time to penetrate further, and to see what happens in the very soul of the mathematician. For this purpose I think I cannot do better than recount my personal recollections. Only I am going to confine myself to relating how I wrote my first treatise on Fuchsian functions. I must apologize, for I am going to introduce some technical expressions, but they need not alarm the reader, for he has no need to under- stand them. I shall say, for instance, that I found the demonstration of such and such a theorem under such and such circumstances; the theorem will have a barbarous name that many will not know, but that is of no importance. What is interesting for the psychologist is not the theorem but the circumstances.

For a fortnight I had been attempting to prove that there could not be any function analogous to what I have since called Fuchsian functions. I was at that time very ignorant. Every day I sat down at my table and spent an hour or two trying a great number of combinations, and I arrived at no result. One night I took some black coffee, contrary to my custom, and was unable to sleep. A host of ideas kept surging in my head; I could almost feel then jostling one another, until two of them coalesced, so to speak, to form a stable combination. When morning came, I had established the existence of one class of Fuchsian functions, those that are derived from the hyper-geometric series. I had only to verify the results, which only took a few hours.

Then I wished to represent these functions by the quotient of two series. This idea was perfectly conscious and deliberate; I was guided by the analogy with elliptical functions. I asked myself what must be the properties of these series, if they existed, and I succeeded without difficulty in forming the series that I have called Theta-Fuchsian.

At this moment I left Caen, where I was then living, to take part in a geological conference arranged by the School of Mines. The incidents of the journey made me forget my mathematical work. When we arrived at Coutances, we got into a break to go for a drive, and, just as I put my foot on the step, the idea came to me, though nothing in my former thoughts seemed to have prepared me for it, that the transformations I had used to define Fuchsian functions were identical with those of non-Euclidian geometry. I made no verification, and had no time to do so, since I took up the conversation again as soon as I had sat down in the break, but I felt absolute certainty at once. When I got back to Caen I verified the result at my leisure to satisfy my conscience.

I then began to study arithmetical questions without any great apparent result, and without suspecting that they could have the least connexion with my previous researches. Disgusted at my want of success, I went away to spend a few days at the seaside, and thought of entirely different things. One day, as I was walking on the cliff, the idea came to me, again with the same characteristics of conciseness, suddenness, and immediate certainty, that arithmetical transformations of indefinite ternary quadratic forms are identical with those of non-Euclidian geometry.

Returning to Caen, I reflected on this result and deduced its consequences. The example of quadratic forms showed me that there are Fuchsian groups other than those which correspond with the hypergeometric series; I saw that I could apply to them the theory of the Theta-Fuchsian series, and that, consequently, there are Fuchsian functions other than those which are derived from the hypergeometric series, the only ones I knew up to that time. Naturally, I proposed to form all these functions. I laid siege to them systematically and captured all the outworks one after the other. There was one, however, which still held out, whose fall would carry with it that of the central fortress. But all my efforts were of no avail at first, except to make me better understand the difficulty, which was already something. All this work was perfectly conscious.

Thereupon I left for Mont-Valerien, where I had to serve my time in the army, and so my mind was preoccupied with very different matters. One day, as I was crossing the street, the solution of the difficulty which had brought me to a standstill came to me all at once. I did not try to fathom it immediately, and it was only after my service was finished that I returned to the question. I had all the elements, and had only to assemble and arrange them. Accordingly I composed my definitive treatise at a sitting and without any difficulty.

It is useless to multiply examples, and I will content myself with this one alone. As regards my other researches, the accounts I should give would be exactly similar, and the observations related by other mathematicians in the enquiry of *l'Enseignement Mathématique* would only confirm them.

One is at once struck by these appearances of sudden illumination, obvious indications of a long course of previous unconscious work. The part played by this unconscious work in mathematical discovery seems to me indisputable, and we shall find traces of it in other cases where it is less evident. Often when a man is working at a difficult question, he accomplishes nothing the first time he sets to work.

Then he takes more or less of a rest, and sits down again at his table. During the first half-hour he still finds nothing, and then all at once the decisive idea presents itself to his mind. We might say that the conscious work proved more fruitful because it was interrupted and the rest restored force and freshness to the mind. But it is more probable that the rest was occupied with unconscious work, and that the result of this work was afterwards revealed to the geometrician exactly as in the cases I have quoted, except that the revelation, instead of coming to light during a walk or a journey, came during a period of conscious work, but independently of that work, which at most only performs the unlocking process, as if it were the spur that excited into conscious form the results already acquired during the rest, which till then remained unconscious.

There is another remark to be made regarding the conditions of this unconscious work, which is, that it is not possible, or in any case not fruitful, unless it is first preceded and then followed by a period of conscious work. These sudden inspirations are never produced (and this is sufficiently proved already by the examples I have quoted) except after some days of voluntary efforts which appeared absolutely fruitless, in which one thought one had accomplished nothing, and seemed to be on a totally wrong track. These efforts, however, were not as barren as one thought; they set the unconscious machine in motion, and without them it would not have worked at all, and would not have produced anything.

The necessity for the second period of conscious work can be even more readily understood. It is necessary to work out the results of the inspiration, to deduce the immediate consequences and put them in order and to set out the demonstrations; but, above all, it is necessary to verify them. I have spoken of the feeling of absolute certainty which accompanies the inspiration; in the cases quoted this feeling was not deceptive, and more often than not this will be the case. But we must beware of thinking that this is a rule without exceptions. Often the feeling deceives us without being any less distinct on that account, and we only detect it when we attempt to establish the demonstration. I have observed this fact most notably with regard to ideas that have come to me in the morning or at night when I have been in bed in a semi-somnolent condition.

Such are the facts of the case, and they suggest the following reflections. The result of all that precedes is to show that the unconscious ego, or, as it is called, the subliminal ego, plays a most important part in mathematical discovery. But the subliminal ego is generally thought of as purely automatic. Now we have seen that mathematical work is not a simple mechanical work, and that it could not be entrusted to any machine, whatever the degree of perfection we suppose it to have been brought to. It is not merely a question of applying certain rules, of manufacturing as many combinations as possible according to certain fixed laws. The combinations so obtained would be extremely numerous, useless, and encumbering.

The real work of the discoverer consists in choosing between these combinations with a view to eliminating those that are useless, or rather not giving himself the trouble of making them at all. The rules which must guide this choice are extremely subtle and delicate, and it is practically impossible to state them in precise language; they must be felt rather than formulated. Under these conditions, how can we imagine a sieve capable of applying them mechanically?

The following, then, presents itself as a first hypothesis. The subliminal ego is in no way inferior to the conscious ego; it is not purely automatic; it is capable of discernment; it has tact and lightness of touch; it can select, and it can divine. More than that, it can divine better than the conscious ego, since it succeeds where the latter fails. In a word, is not the subliminal ego superior to the conscious ego? The importance of this question will be readily understood. In a recent lecture, M. Boutroux showed how it had arisen on entirely different occasions, and what consequences would be involved by an answer in the affirmative. (See also the same author's *Science et Religion*, pp. 313 *et seq.*)

Are we forced to give this affirmative answer by the facts I have just stated? I confess that, for my part, I should be loth to accept it. Let us, then, return to the facts, and see if they do not admit of some other explanation.

It is certain that the combinations which present themselves to the mind in a kind of sudden illumination after a somewhat prolonged period of unconscious work are generally useful and fruitful combinations, which appear to be the result of a preliminary sifting. Does it follow from this that the subliminal ego, having divined by a delicate intuition that these combinations could be useful, has formed none but these, or has it formed a great many others which were devoid of interest, and remained unconscious?

Under this second aspect, all the combinations are formed as a result of the automatic action of the subliminal ego, but those only which are interesting find their way into the field of consciousness. This, too, is most mysterious. How can we explain the fact that, of the thousand products of our unconscious activity, some are invited to cross the threshold, while others remain outside? Is it mere chance that gives them this privilege? Evidently not. For instance, of all the excitements of our senses, it is only the most intense that retain our attention, unless it has been directed upon them by other causes. More commonly the privileged unconscious phenomena, those that are capable of becoming conscious, are those which, directly or indirectly, most deeply affect our sensibility.

It may appear surprising that sensibility should be introduced in connexion with mathematical demonstrations, which, it would seem, can only interest the intellect. But not if we bear in mind the feeling of mathematical beauty, of the harmony of numbers and forms and of geometric elegance. It is a real aesthetic feeling that all true mathematicians recognize, and this is truly sensibility.

Now, what are the mathematical entities to which we attribute this character of beauty and elegance, which are capable of developing in us a kind of aesthetic emotion? Those whose elements are harmoniously arranged so that the mind can, without effort, take in the whole without neglecting the details. This harmony is at once a satisfaction to our aesthetic requirements, and an assistance to the mind which it supports and guides. At the same time, by setting before our eyes a well-ordered whole, it gives us a presentiment of a mathematical law. Now, as I have said above, the only mathematical facts worthy of retaining our attention and capable of being useful are those which can make us acquainted with a mathematical law. Accordingly we arrive at the following conclusion. The useful combinations are precisely the most beautiful, I mean those that can most charm that special sensibility that all mathematicians know, but of which laymen are so ignorant that they are often tempted to smile at it.

What follows, then? Of the very large number of combinations which the subliminal ego blindly forms, almost all are without interest and without utility. But, for that very reason, they are without action on the aesthetic sensibility; the consciousness will never know them. A few only are harmonious, and consequently at once useful and beautiful, and they will be capable of affecting the geometrician's special sensibility 1 have been speaking of; which, once aroused, will direct our attention upon them, and will thus give them the opportunity of becoming conscious.

This is only a hypothesis, and yet there is an observation which tends to confirm it. When a sudden illumination invades the mathematician's mind, it most frequently happens that it does not mislead him. But it also happens sometimes, as I have said, that it will not stand the test of verification. Well, it is to be observed almost always that this false idea, if it had been correct, would have flattered our natural instinct for mathematical elegance.

Thus it is this special aesthetic sensibility that plays the part of the delicate sieve of which I spoke above, and this makes it sufficiently clear why the man who has it not will never be a real discoverer.

All the difficulties, however, have not disappeared. The conscious ego is strictly limited, but as regards the subliminal ego, we do not know its limitations, and that is why we are not too loth to suppose that in a brief space of time it can form more different combinations than could be comprised in the whole life of a conscient being. These limitations do exist, however. Is it conceivable that it can form all the possible combinations, whose number staggers the imagination? Nevertheless this would seem to be necessary, for if it produces only a small portion of the combinations, and that by chance, there will be very small likelihood of the *right* one, the one that must be selected, being found among them.

Perhaps we must look for the explanation in that period of preliminary conscious work which always precedes all fruitful unconscious work. If I may be permitted a crude comparison, let us represent the future elements of our combinations as something resembling Epicurus's hooked atoms. When the mind is in complete repose these atoms are immovable; they are, so to speak, attached to the wall. This complete repose may continue indefinitely without the atoms meeting, and, consequently, without the possibility of the formation of any combination.

On the other hand, during a period of apparent repose, but of unconscious work, some of them are detached from the wall and set in motion. They plough through space in all directions, like a swarm of gnats, for instance, or, if we prefer a more learned comparison, like the gaseous molecules in the kinetic theory of gases. Their mutual collisions may then produce new combinations.

What is the part to be played by the preliminary conscious work? Clearly it is to liberate some of these atoms, to detach them from the wall and set them in motion. We think we have accomplished nothing, when we have stirred up the elements in a thousand different ways to try to arrange them, and have not succeeded in finding a satisfactory arrangement. But after this agitation imparted to them by our will, they do not return to their original repose, but continue to circulate freely.

Now our will did not select them at random, but in pursuit of a perfectly definite aim. Those it has liberated are not, therefore, chance atoms; they are those from which we may reasonably expect the desired solution. The liberated atoms will then experience collisions, either with each other, or with the atoms that have remained stationary, which they will run against in their course. I apologize once more. My comparison is very crude, but I cannot well see how I could explain my thought in any other way.

However it be, the only combinations that have any chance of being formed are those in which one at least of the elements is one of the atoms deliberately selected by our will. Now it is evidently among these that what I called just now the *right* combination is to be found. Perhaps there is here a means of modifying what was paradoxical in the original hypothesis.

Yet another observation. It never happens that unconscious work supplies *ready-made* the result of a lengthy calculation in which we have only to apply fixed rules. It might be supposed that the subliminal ego, purely automatic as it is, was peculiarly fitted for this kind of work, which is, in a sense, exclusively mechanical. It would seem that, by thinking overnight of the factors of a multiplication sum, we might hope to find the product ready-made for us on waking; or, again, that an algebraical calculation, for instance, or a verification could be made unconsciously. Observation proves that such is by no means the case. All that we can hope from these inspirations, which are the fruits of unconscious work, is to obtain points of departure for such calculations. As for the calculations themselves, they must be made in the second period of conscious work which follows the inspiration, and in which the results of the inspiration are verified and the consequences deduced. The rules of these calculations are strict and complicated; they demand discipline, attention, will, and consequently consciousness. In the subliminal ego, on the contrary, there reigns what I would call liberty, if one could give this name to the mere absence of discipline and to disorder born of chance. Only, this very disorder permits of unexpected couplings.

I will make one last remark. When I related above some personal observations, I spoke of a night of excitement, on which I worked as though in spite of myself The cases of this are frequent, and it is not necessary that the abnormal cerebral activity should be caused by a physical stimulant, as in the case quoted. Well, it appears that, in these cases, we are ourselves assisting at our own unconscious work, which becomes partly perceptible to the overexcited consciousness, but does not on that account change its nature. We then become vaguely aware of what distinguishes the two mechanisms, or, if you will, of the methods of working of the two egos. The psychological observations I have thus succeeded in making appear to me, in their general characteristics, to confirm the views I have been enunciating.

Truly there is great need of this, for in spite of everything they are and remain largely hypothetical. The interest of the question is so great that I do not regret having submitted them to the reader.

# Endnotes

1. Born in 1926, winner of both the Fields Medal and the Abel Prize and a member of the French Academy of Sciences, Jean-Pierre Serres is considered one of the greatest mathematicians of the twentieth century.
2. Yitang Zhang, a Chinese-born American mathematician (b. 1955), achieved international renown in 2013 for his pathbreaking work on the distribution of prime numbers. Roger Apéry (1916–94), a French mathematician, demonstrated in 1977 the irrationality of $\zeta(3)$, the value of the Riemann zeta function at three, subsequently named Apéry's constant in his honor.
3. In "Geometry and Experience," a lecture delivered to the Prussian Academy of Sciences in January 1921, Einstein remarked: "One reason why mathematics enjoys special esteem, above all other sciences, is that its laws are absolutely certain and indisputable, while those of other sciences are to some extent debatable and in constant danger of being overthrown by newly discovered facts"; in *Sidelights on Relativity,* trans. G. B. Jeffery and W. Perrett (London: Methuen, 1922), 27.
4. From the Senagalese mathematician Hamet Seydi, former dean of the faculty of sciences and technology at Cheikh Anta Diop University in Dakar and a collaborator of Alexandre Grothendieck, I learned the origin of Senghor's phrase. Senghor, then President of the Republic, was to give the opening address of an international mathematics conference in Dakar. When Seydi showed him the conference program, Senghor contemplated the mysterious titles of the lectures in silence, and then exclaimed, "What you mathematicians do is the poetry of the sciences!"
5. One enthusiastic and cultivated reader wrote me to say, "Your prose, sir, is post-Mallarméan and pre-Symbolist." Or perhaps "pre-hermeneutic"? See my *Birth of a Theorem: A Mathematical Adventure,* trans. Malcolm DeBevoise (New York: Farrar, Straus & Giroux, 2015); originally published as *Théorème vivant* (Paris: Grasset, 2012).
6. Consider, for example, the title of one of Seydi's own articles: "On the Theory of Excellent Rings in Characteristic Zero." The technical meaning is very precise, but might one not mistake it for an Oulipian verse?

7. The works inspired by Man Ray's visits to the IHP between 1934 and 1936 are described by Isabelle Fortuné, "Man Ray et les objets mathématiques," *Études photographiques* 6 (May 1999); https://journals.openedition.org/etudesphotographiques/190. See also Wendy A. Grossman and Édouard Sebline, eds, *Man Ray—Human Equations: A Journey from Mathematics to Shakespeare* (Ostfildern, Germany: Hatje Cantz, 2015). For a broader perspective on geometric models, see Jean-Philippe Uzan and Cédric Villani, eds, *Objets mathématiques* (Paris: CNRS Éditions, 2017).

8. To take just one example: the White Knight's explanation to Alice, in *Through the Looking Glass*, of the difference between a name and the name of a name cannot help but call to mind, for a computer programmmer, the distinction between the address of a pointer and that of the referenced object; for a mathematician, the difference between a set and the property that defines it; and for a logician, Gödel's discussion of the numbering of propositions in the proof of his famous incompleteness theorem.

9. See my filmed lecture "Pour faire naître une idée," in *Cédric Villani: Un mathématicien aux métallos*, 4-DVD boxed set, ARTE Éditions, 2017.

10. A shorthand for Ouvroir de littérature potentielle (Workshop of Potential Literature).—Trans.

11. Thus the title of Queneau's book *Cent milles millards de poèmes* (Paris: Gallimard, 1961).

12. Raymond Queneau, in Oulipo, *La Littérature potentielle: Créations, re-créations, récréations* (Paris: Gallimard, 1973).

13. La Fontaine's lines, translated by many hands over the past three centuries, are as well known in English as they are in French. In Elizur Wright's famous 1841 version they read:

> A Grasshopper gay
> Sang the summer away,
> And found herself poor
> By the winter's first roar.
> Of meat or of bread,
> Not a morsel she had!
> So a begging she went,
> To her neighbour the ant....

A literal translation from the Oulipian version yields only the purest nonsense (the fifth line, for example, becomes "Not a sexed petrographic

Camberwell beauty of skunk or studboar"), without preserving simi-
larities of sound or spelling. A comparable English version could be
obtained only by applying the S + 7 rule to an English translation.—
Trans.

14. A shorthand for Ouvroir de bande dessinée potentielle (Workshop of
    Potential Comic Book Art).—Trans.

15. See Boris Vian, *Mémoire concernant le calcul numérique de Dieu par des méthodes
    simples et fausses* [1955], privately published in 1977; conserved by the
    Bibliothèque nationale de France, Arsenal, 4-Z-6571.

16. See Chapter 15 of my *Birth of a Theorem*, 86–94.

17. See Cédric Villani, "Paradoxe de Scheffer-Shnirelman revue sous
    l'angle de l'intégration convex [d'après C. De Lellis et L. Székelyhidi],"
    November 2008 Bourbaki Seminar talk, no. 1001; https://cedricvillani.
    org/sites/dev/files/old_images/2012/08/B10.Bourbaki2.pdf.

18. See Chapter 25 of my *Birth of a Theorem*, 134–8.

19. Henri Poincaré, *Science and Method* [1908], trans. Francis Maitland (London:
    T. Nelson, 1914), 1.2, 34.

20. A common (and not inaccurate) paraphrase of Einstein's actual
    words, from a 1936 article; see Alice Calaprice, ed., *The Expanded Quotable
    Einstein* (Princeton: Princeton University Press, 2000), 278.—Trans.

21. Attempts have sometimes been made, much less successfully, it must
    be said, to win acceptance for mathematics as the language of the
    human sciences.

22. See, for example, Cédric Villani, "L'écriture des mathématiciens," in
    Éric Guichard, ed., *Écritures: Sur les traces de Jack Goody* (Lyon: Presses de
    Enssib, 2012); available at http://barthes.ens.fr/articles/Villani-ecriture-
    mathematiciens.pdf.

23. See S. J. Szarek and D. Voiculescu [2000], "Shannon's Entropy Power
    Inequality via Restricted Minkowski Sums," in V. D. Milman and
    G. Schechtman, eds, *Geometric Aspects of Functional Analysis,* Lecture Notes
    in Mathematics, vol. 1745 (Berlin: Springer, 2007), 257–62.

24. See A. Grothendieck, "Classes de faisceaux et théorème de Riemann-
    Roch" [1957]; published in *Séminaire de géométrie algébrique* [SGA 6], Lecture
    Notes in Mathematics 225 (Berlin: Springer, 1971).

25. From a letter to Madame Schabelskoy; a slightly different version is
    found in *Sónya Kovalévsky: Her Recollections of Childhood,* trans. Isabel F. Hapgood
    (New York: Century Co., 1895), 316.—Trans.

26. On the Poincaré Conjecture and Perelman's proof of it, see Masha Gessen, *Perfect Rigor: A Genius and the Mathematical Breakthrough of the Century* (Boston: Houghton Mifflin Harcourt, 2009).

27. Grothendieck's unpublished manuscript *Récoltes et semailles: Réflexions et témoignage sur un passé d'un mathématicien* (1986) is now available in PDF format via https://uberty.org/wp-content/uploads/2015/12/Grothendeick-RetS.pdf.

28. I have been complimented many times for the fresh and informal style of my first book book, *Topics in Optimal Transportation* (Providence, RI: American Mathematics Society, 2003). But now, more than a decade later, I wonder how I could have allowed myself to write in such a relaxed manner!

29. See Poincaré, *Science and Method*, 1.3, 46–63.

30. Readers who are pressed for time may content themselves with reading the heart of Poincaré's essay, the famous passage in which the inspiration needed to master Fuchsian functions comes from drinking a cup of black coffee one evening, from stepping onto a horse-drawn carriage (an "omnibus") in Coutances, from walking along a cliff above the sea near Caen, and from crossing a street in Saint-Valérian (pp. 51–53 below). Those who have the leisure to read the entire essay will be rewarded by the pleasure of reading a mathematician's delightful prose at greater length.

31. These are the opening lines of my contribution to the series "Obsession du mois" for October 2012 in the *Nouvel Observateur*; https://o.nouvelobs.com/pop-life/20121008.OBS4885/obsession-du-mois-cedric-villani.html.

32. See my filmed lectures in the DVD set *Cédric Villani: Un mathématicien aux métallos.*

33. The festival was founded in 1999; for the 2012 program, see the archive at http://www.lamilanesiana.eu/.

34. Henri Poincaré, *The Value of Science* [1905], in *The Foundations of Science,* trans. George Bruce Halsted (New York: Science Press, 1913); reprinted as a separate volume in 1958 by Dover Press.

35. Poincaré, *Science and Method*, 1.4, 66.

36. See section 5 of the prologue to Friedrich Nietzsche, *Thus Spoke Zarathustra*, in *The Portable Nietzsche*, ed. and trans. Walter Kaufmann (New York: Viking, 1954), 129.

37. Voltaire, *Éléments de la philosophie de Newton* (Amsterdam, 1738), part 3, chapter 5; in *Les Œuvres complètes de Voltaire,* ed. Theodore Besterman et al., 200 vols (Oxford: Voltaire Foundation), 15:435–6.

38. I describe Nash's achievements at greater length in Chapter 29 of *Birth of a Theorem*, 166–70.

39. See John Nash, "$C^1$-isometric imbeddings," *Ann. Math.* 60, no. 2 (1954): 383–96; and "The imbedding problem for Riemannian manifolds," *Ann. Math.* 63, no. 1 (1956): 20–63.

40. See Sylvia Nasar, *A Beautiful Mind: A Biography of John Forbes Nash, Jr.* (New York: Simon and Schuster, 1998), 155–163. For a more technical discussion, see my article "Nash et les équations aux dérivées partielles," *Matapli108* (4 November 2015): 35–53; https://cedricvillani.org/sites/dev/files/old_images/2015/12/Matapli108_C_Villani.pdf.

# Index